Captivating Cells Book

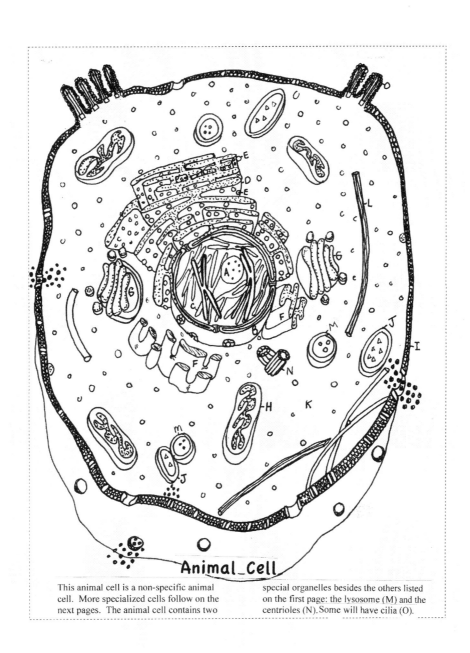

Animal_Cell

This animal cell is a non-specific animal cell. More specialized cells follow on the next pages. The animal cell contains two special organelles besides the others listed on the first page: the lysosome (M) and the centrioles (N). Some will have cilia (O).

By Barbara Sabet

Table of Contents

Captivating Cells Coloring Book

Introduction

What is a cell? A cell is the smallest living unit. All living things are made of cells, and come from cells. Even the lowly bacteria are cells. The cell unit is so well-organized that it can exist all by itself in certain organisms. Most cells have the same basic parts (organelles) and some cells have extra organelles to help them do a specific job.

This book will show you 30 different cells from many different organisms. Some are found in the human body and some live all by themselves! the cells found in the human body often work with other cells of the same type to form a tissue. Tissues can then be shaped to form organs like a stomach or heart.

The drawings are labeled with letters. The common organelles to all cells except bacteria are labeled below. When a cell has a part that is special to that cell, it is listed below the drawing.

Feel free to color the parts any way you like, since in real life, cells are mostly colorless except for a few cells. Most plant cells are green because they are full of a green pigment called chlorophyll that is found in their organelles called "chloroplasts." It would be a good idea to color the chloroplasts green. Red blood cells are red because they contain a lot of iron which has a reddish tinge. You should color the entire Red blood cell a red color.

Labeling

A– **Nucleolus:** Found in the nucleus, it makes the parts of a ribosome (protein factory)

B– **Nucleus:** It is the control center of the cell and contains our hereditary structures called chromosomes.

C– **Chromosome:** Found in the nucleus, it contains DNA and controls hereditary

D– **Endoplasmic Reticulum (E.R.):** It looks like ribbon, and is the transportation network of the cell. The rough ER has ribosomes and the smooth ER (letter F) does not.

E– **Ribosome:** Can be found on the E.R. or alone in the cytoplasm. It makes proteins.

F– **Smooth ER:** Does not contain ribosomes, but is also for transportation.

G– **Golgi body:** It looks like a stack of pancakes, and it packages proteins made by ribosomes.

H– **Mitochondria:** They look like peanuts, and there job is to make energy for cellular activities.

I– **Plasma membrane:** It forms a barrier around the cell and has tiny little openings in it to let some things in and out of the cell. It is particular about what can travel across it.

J–**Vacuole:** These organelles are for storing different substances

K–**Cytoplasm:** This is the background "gel" of the cell that all the organelles swim around in.

L– **Cytoskeleton:** A system of different sized tubes that provide support for the cell.

Animal Cells

Even though animal cells are very diverse, with many different shapes and purposes, they have many things in common. They all have a nucleus, chromosomes, endoplasmic reticulum, golgi bodies, mitochondria, lysosome, centriole, ribosomes, vacuoles, plasma membranes (cell membranes) and a cytoskeleton. They do not have cell walls or chloroplasts.

If you look at some of the animal cells in this book, you might begin to think that their *structure determines their function.* In other words, what a structure can do depends on it's form. We see this is true in not just cells, but in organs too. Look at the shape of the goblet cell that is found lining the respiratory tract in animals. Since it's job is to make mucous that traps dirt and other particles so we can cough them out, it is shaped like a big wine goblet, full of mucous. It has more E.R. and golgi bodies than the average cell so that it can produce lots of mucous and package it. All the other organelles are pressed to the sides to make the most room for the precious mucous!

Look at the shape of the neuron (nerve cell). Why does it seem to have a tail? Well, since it's job is to send impulses to the next neuron, the tail (which is covered in a fatty substance that acts as an insulator) sends those impulses very quickly to the next neuron. The dendrites on the body of the neuron are made in such a way as to accept impulses from the neurons before them. Everything happens quickly and efficiently.

The adipose cell (fat cell) is basically just a big bag of fat which can expand and contract depending on the calories the animal consumes. The cell has all the normal organelles, but has a great big vacuole to hold the fat. Have fun coloring them!

Animal cells live with other animal cells, and so are animals are considered *multicellular,* not *unicellular* like protists, bacteria and some fungi.

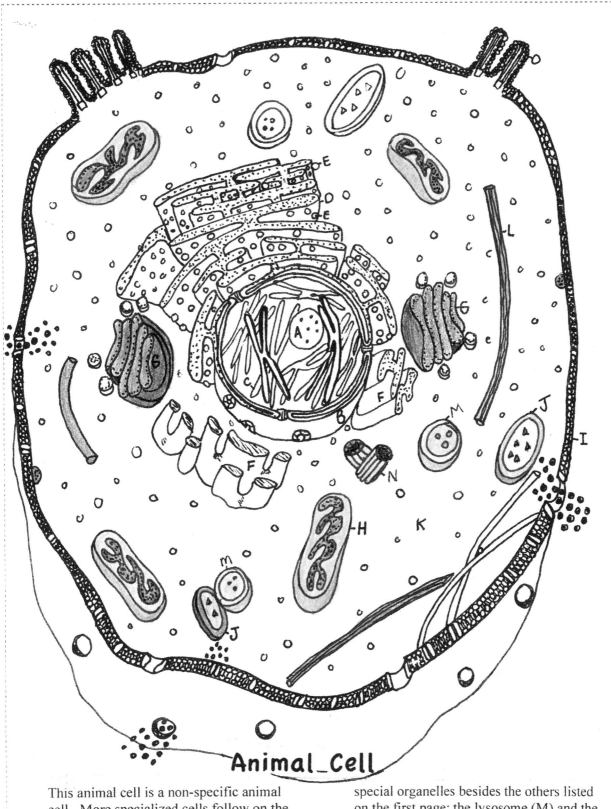

Animal_Cell

This animal cell is a non-specific animal cell. More specialized cells follow on the next pages. The animal cell contains two special organelles besides the others listed on the first page: the lysosome (M) and the centrioles (N). Some will have cilia (O).

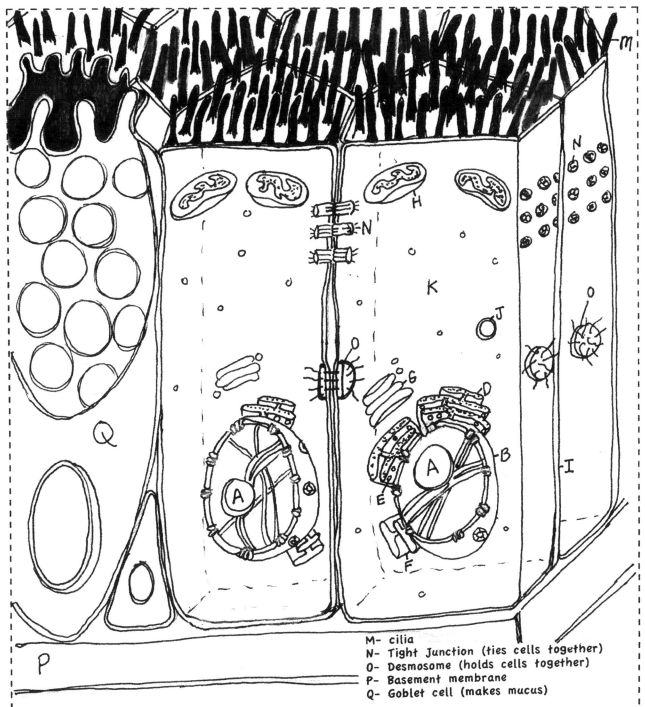

M- cilia
N- Tight Junction (ties cells together)
O- Desmosome (holds cells together)
P- Basement membrane
Q- Goblet cell (makes mucus)

Ciliated Columnar Epithelial Cell

This cell is found lining the respiratory tract (lungs and airways) and the fallopian tubes in the female. They have around 200 cilia that protrude from the top of the cell. The cilia help to move mucus and dirt away from the lungs and in the fallopian tubes, help to move the egg along towards the uterus. They are usually found with goblet cells between them (form mucus).

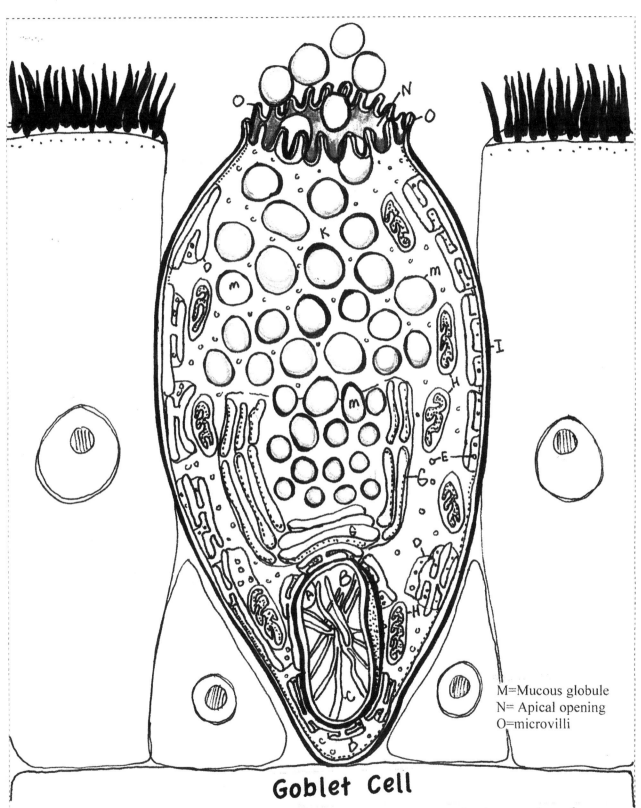

M=Mucous globule
N= Apical opening
O=microvilli

Goblet Cell

The goblet cell is found in places where mucous is needed as in the respiratory tract and digestive tract. The golgi body is unusually large so that it can package the mucous in large droplets for transport out of the cell.

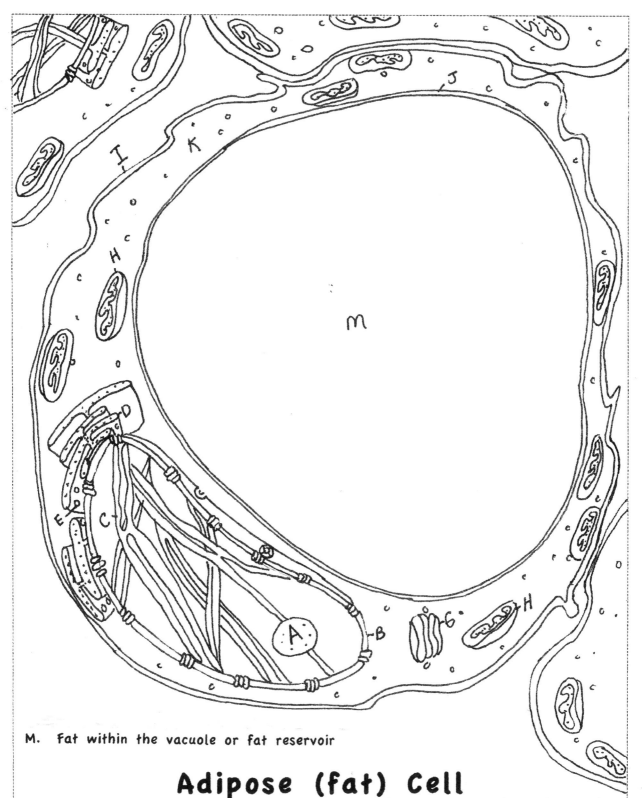

M. Fat within the vacuole or fat reservoir

Adipose (fat) Cell

The fat cell is part of the adipose tissue that is found throughout the body under the skin and around the organs. It is where energy is stored as fat. Fat cells have a large vacuole that can fill up and empty itself of fat. So when a person loses weight, the vacuole loses fat, the cell gets smaller, and so does the person. When a person gains weight, the vacuole accepts more fat up to a certain point.

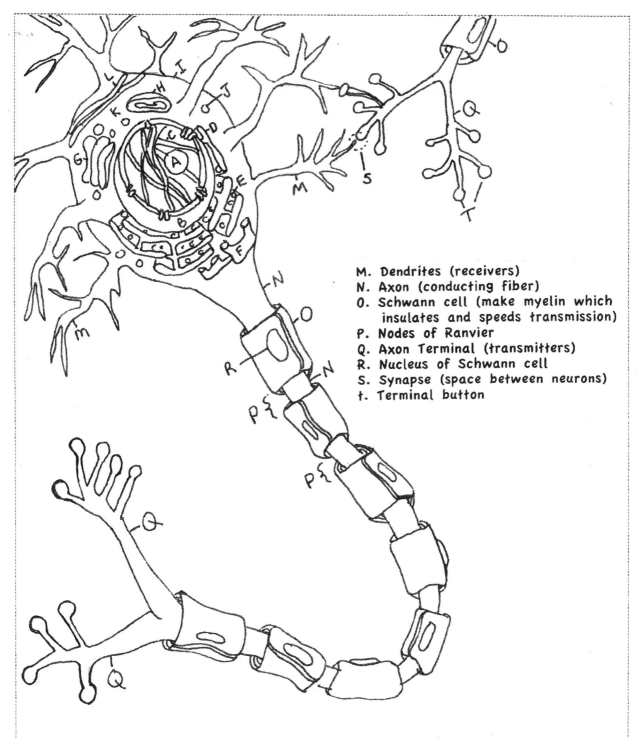

M. Dendrites (receivers)
N. Axon (conducting fiber)
O. Schwann cell (make myelin which insulates and speeds transmission)
P. Nodes of Ranvier
Q. Axon Terminal (transmitters)
R. Nucleus of Schwann cell
S. Synapse (space between neurons)
t. Terminal button

Nerve Cell (Neuron)

Nerve cells are also called neurons. They are found throughout the body. Nerves go to and from the spinal cord as they go to and from organs and body parts. There are many neurons in the brain. Neurons have a special shape which includes a long tail (axon) which carries the nerve impulse down the neuron and branching structures on the cell called dendrites which connect to other neurons.

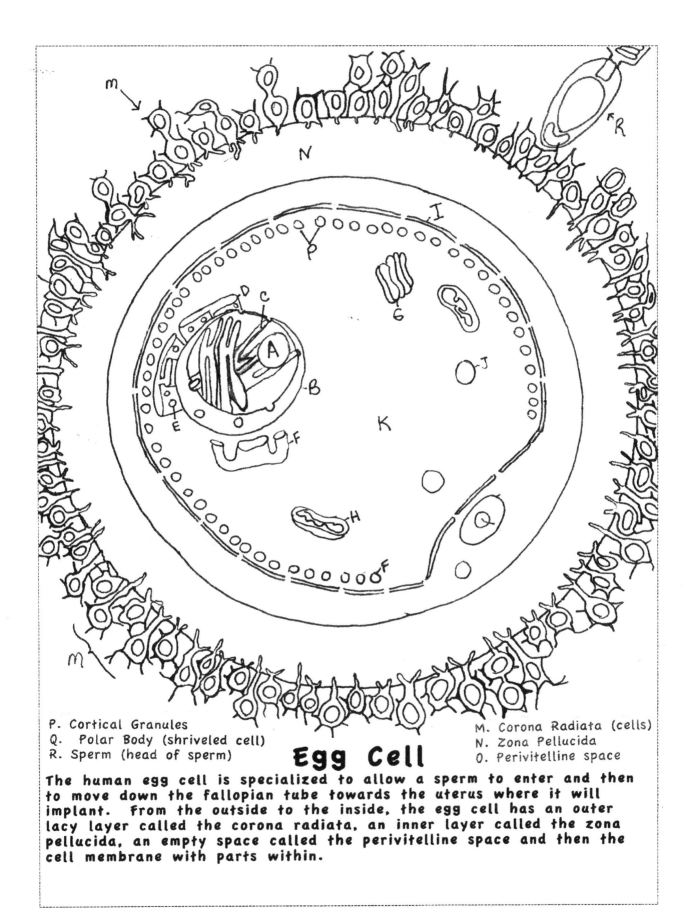

P. Cortical Granules
Q. Polar Body (shriveled cell)
R. Sperm (head of sperm)

M. Corona Radiata (cells)
N. Zona Pellucida
O. Perivitelline space

Egg Cell

The human egg cell is specialized to allow a sperm to enter and then to move down the fallopian tube towards the uterus where it will implant. From the outside to the inside, the egg cell has an outer lacy layer called the corona radiata, an inner layer called the zona pellucida, an empty space called the perivitelline space and then the cell membrane with parts within.

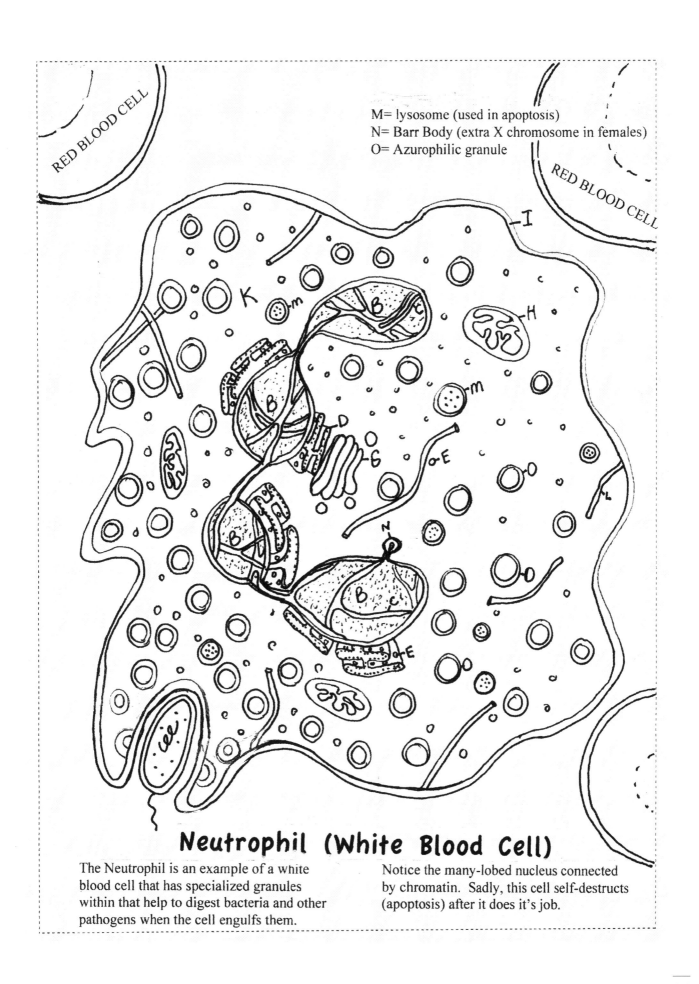

M= lysosome (used in apoptosis)
N= Barr Body (extra X chromosome in females)
O= Azurophilic granule

RED BLOOD CELL

RED BLOOD CELL

Neutrophil (White Blood Cell)

The Neutrophil is an example of a white blood cell that has specialized granules within that help to digest bacteria and other pathogens when the cell engulfs them.

Notice the many-lobed nucleus connected by chromatin. Sadly, this cell self-destructs (apoptosis) after it does it's job.

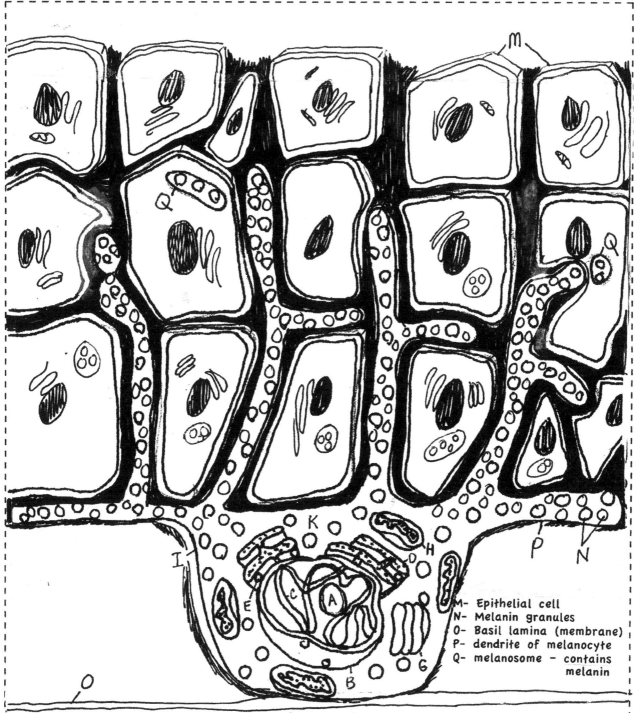

M— Epithelial cell
N— Melanin granules
O— Basil lamina (membrane)
P— dendrite of melanocyte
Q— melanosome – contains melanin

Melanocyte

This cell produces melaqnin which is responsible for skin color. The cell is found in the bottom layer of the epidermis of the skin and is also found in eyes and hair. Melanin is reponsible for absorbing all of the UV-B light so that it cannot pass the skin layer into the layers beneath the skin. Lighter skinned people have low levels of melanogenesis (production of melanin). Exposure to sunlight will increase melanogenesis and therefore make the skin darker.

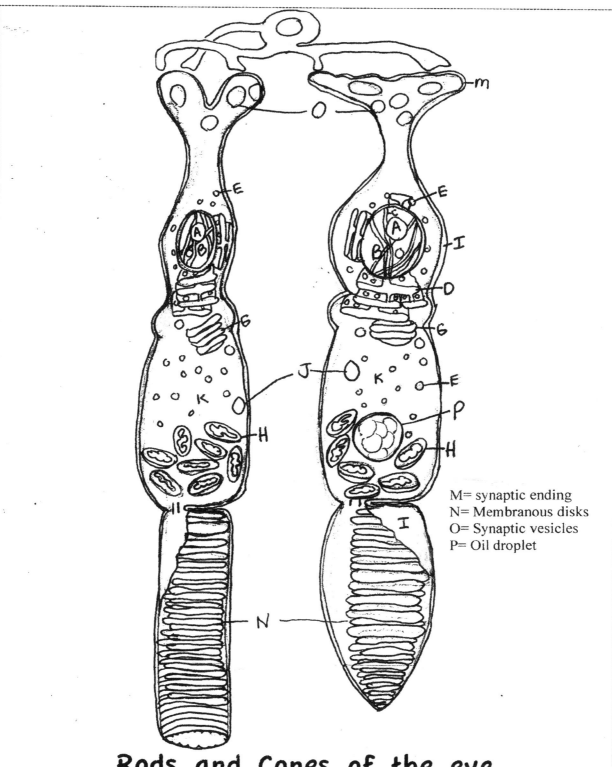

M= synaptic ending
N= Membranous disks
O= Synaptic vesicles
P= Oil droplet

Rods and Cones of the eye

Rods and cones are the two types of photoreceptor cells in the retina of the eye responsible for black and white (rods) and color vision (cones). They have some specialized structures which are membranous disks that contain rhodopsin and photopsin which are proteins that help to see color and black and white.

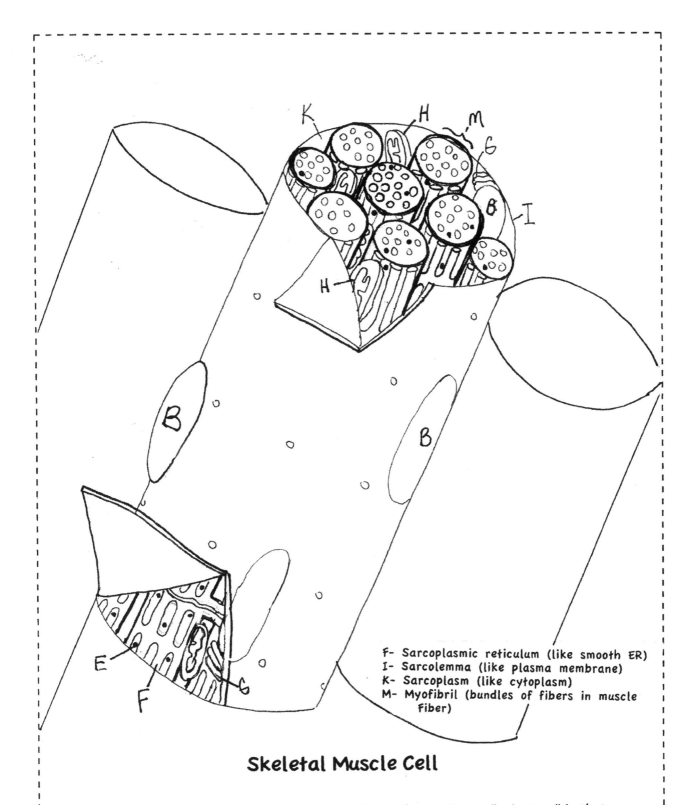

F- Sarcoplasmic reticulum (like smooth ER)
I- Sarcolemma (like plasma membrane)
K- Sarcoplasm (like cytoplasm)
M- Myofibril (bundles of fibers in muscle
fiber)

Skeletal Muscle Cell

These elongated cells are attached to bones by tendons. These cells are "voluntary" in that you
have to volunteer to use them (not automatic like the heart). Sometimes muscle cells are called
muscle fibers. They have many nucleii. Many of these muscle cells or fibers bundled together make
up a muscle. They are composed of smaller fibers called myofibrils.

Protist Cells

The Protist kingdom is sort of a catch-all kingdom in that scientists seem to put organisms here that don't fit anywhere else. Their cells are like animal cells, but not really. They are sometimes like plant cells, but not really. They have some similarities to bacteria cells, but not totally, and they also have a few fungi-like characteristics.

Protist cells have many of the same organelles found in animal and plant cells, but have a few more. They have a cell wall made of cellulose like plants. Some interesting facts about protists are that they are unicellular (just one cell per organism) and they can almost all move on their own with the help of many cilia, a flagella or pseudopods. They show a remarkable variation in cell organization and cell division.

Most protist cells live in a fluid, such as blood, pond water, ocean water, etc. They take in food through a little mouth, often called a gullet. Once the food has entered the little cell, it is digested with the help of digestive enzymes.

Some protists contain chlorophyll and act like moveable plant cells. The green of pond water is often due to small, green protists called Chlamydomonas. In many species of protists, the type of nutrition they get depends on the environment they are in. When there's lots of light, they act like a plant and carry out photosynthesis, but when it's dark, they eat food.

Some protists don't eat or carry out photosynthesis, but just simply absorb nutrients from around them.

Some protists swim around in their fluid while some sort of slink around on the bottom of their watery environment. The amoeba moves around with the help of pseudopods which are *false feet.*

Scientists think the protists probably appeared around 2 billion years ago, so they are very ancient indeed!

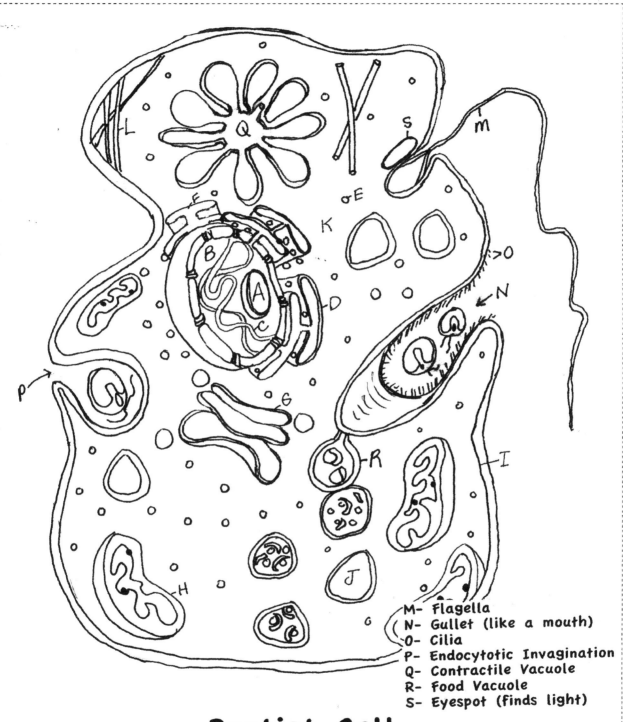

M- Flagella
N- Gullet (like a mouth)
O- Cilia
P- Endocytotic Invagination
Q- Contractile Vacuole
R- Food Vacuole
S- Eyespot (finds light)

Protist Cell

The typical protist cell would have a mouth-like opening to take in food and a way to move on it's own usually through cilia, flagella or pseudopod. It has all the usual organelles and some even have chloroplasts! Most live in water (fresh or salt water) and some even live in blood!

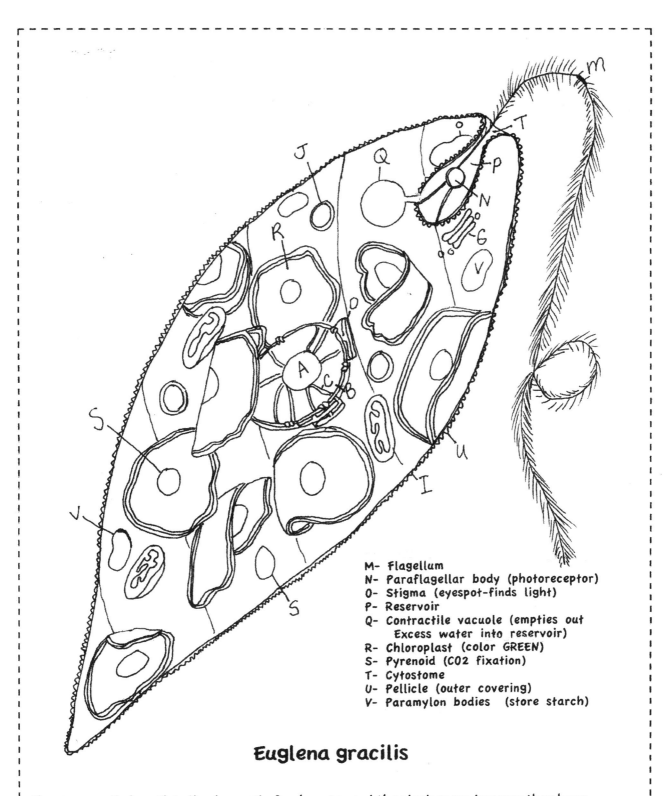

M- Flagellum
N- Paraflagellar body (photoreceptor)
O- Stigma (eyespot-finds light)
P- Reservoir
Q- Contractile vacuole (empties out
    Excess water into reservoir)
R- Chloroplast (color GREEN)
S- Pyrenoid (CO2 fixation)
T- Cytostome
U- Pellicle (outer covering)
V- Paramylon bodies (store starch)

Euglena gracilis

These one-celled protists live in mostly fresh water and they look green because they have chloroplasts within them. Even though they are not plants, they do have chloroplasts like plants do. They have an eyespot that helps them find sunlight since they need the sun to make their own food. But the euglena can also eat food by engulfing it through a process called "phagocytosis."

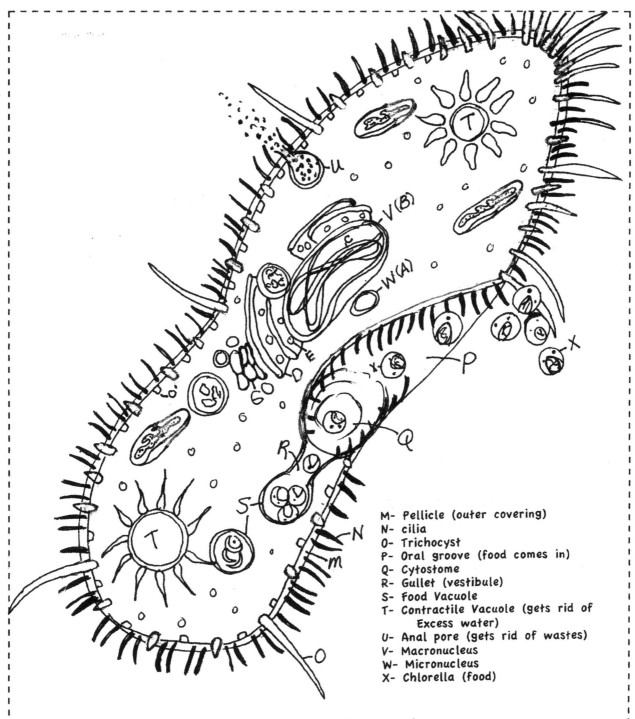

M- Pellicle (outer covering)
N- cilia
O- Trichocyst
P- Oral groove (food comes in)
Q- Cytostome
R- Gullet (vestibule)
S- Food Vacuole
T- Contractile Vacuole (gets rid of
        Excess water)
U- Anal pore (gets rid of wastes)
V- Macronucleus
W- Micronucleus
X- Chlorella (food)

Paramecium bursaria

These one-celled protists live in mostly fresh water but can also be found in marine environments.
They have "cilia" attached to their outer membranes which they use to swim with whiplash
movements. they also use the cilia to sweep in food through their oral groove. They have a shape
similar to a footprintor a cigar. They have an interesting organelle called a "contractile vacuole"
that helps them to expel excess water. Paramecium bursaria forms a symbiotic relationship with
the green algae Zoochlorella which provides food for the paramecium while the paramecium gives
them a place to live.

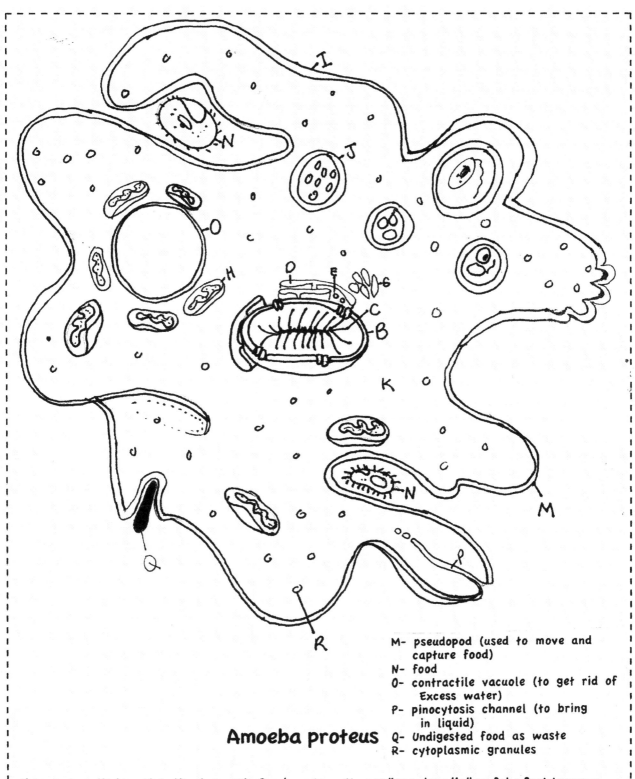

Amoeba proteus

M- pseudopod (used to move and capture food)
N- food
O- contractile vacuole (to get rid of Excess water)
P- pinocytosis channel (to bring in liquid)
Q- Undigested food as waste
R- cytoplasmic granules

These one-celled protists live in mostly fresh water. It uses "pseudopodia" or fake feet to move like a big blob. Due to "phytochromes", this amoeba can be found in a variety of colors (yellow, green and purple). These protists are often studied in science classrooms as they are relatively easy to study under the microscope. They eat smaller organisms and engulf them by phagocytosis.

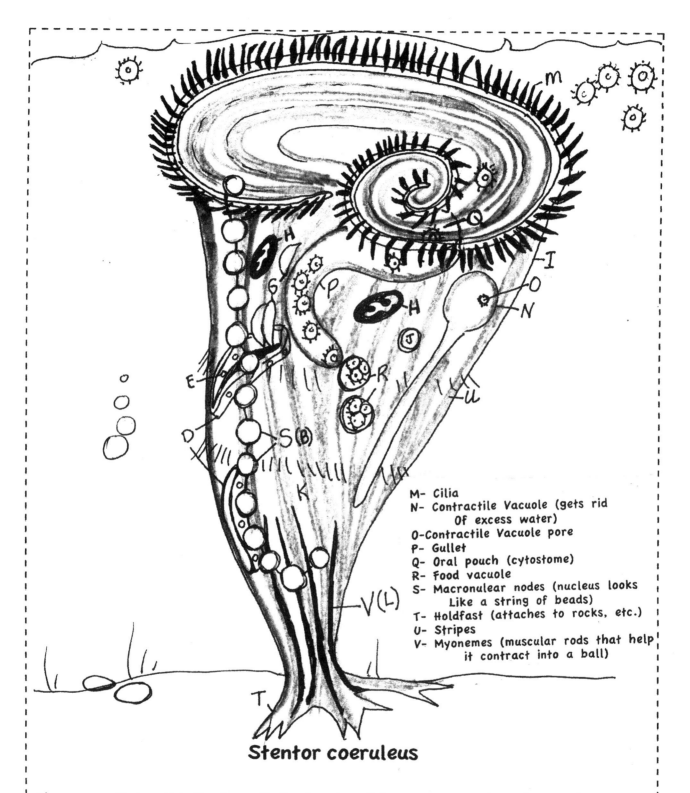

Stentor coeruleus

M- Cilia
N- Contractile Vacuole (gets rid
   Of excess water)
O-Contractile Vacuole pore
P- Gullet
Q- Oral pouch (cytostome)
R- Food vacuole
S- Macronulear nodes (nucleus looks
   Like a string of beads)
T- Holdfast (attaches to rocks, etc.)
U- Stripes
V- Myonemes (muscular rods that help
   it contract into a ball)

These one-celled protists live in mostly fresh water. It has a trumpet or horn shape with a ring of cilia around the bell. These cilia help to sweep in food and also help in swimming. They are among the largest of the protists and come in many different colors. Stentor coeruleus is blue due to the presence of Stentorin, a natural blue pigment. They have a contractile vacuole which helps expel excess water.

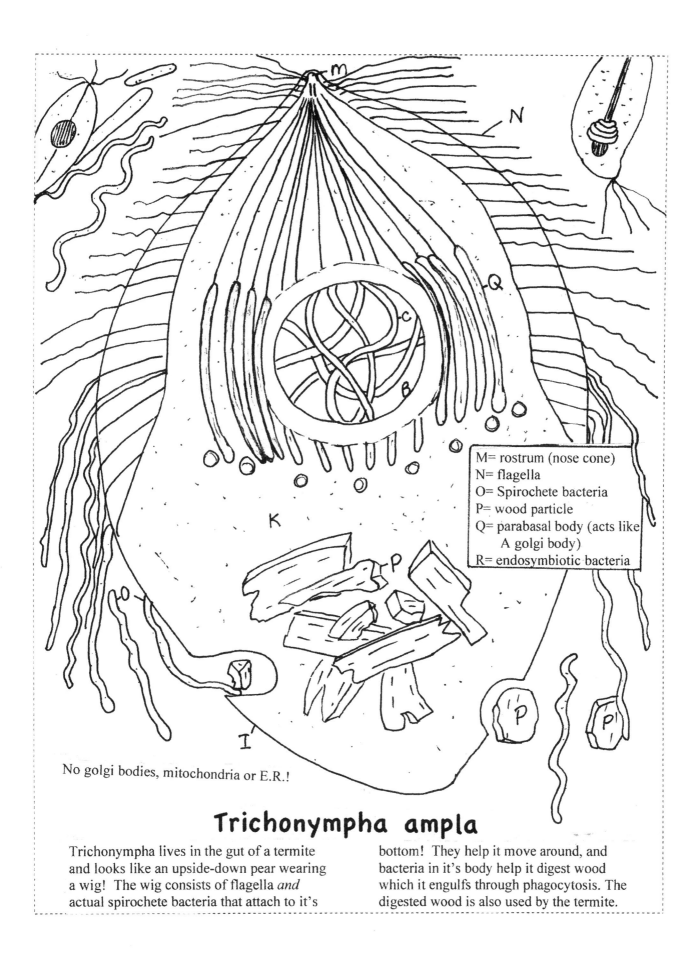

M= rostrum (nose cone)
N= flagella
O= Spirochete bacteria
P= wood particle
Q= parabasal body (acts like
    A golgi body)
R= endosymbiotic bacteria

No golgi bodies, mitochondria or E.R.!

Trichonympha ampla

Trichonympha lives in the gut of a termite and looks like an upside-down pear wearing a wig! The wig consists of flagella *and* actual spirochete bacteria that attach to it's bottom! They help it move around, and bacteria in it's body help it digest wood which it engulfs through phagocytosis. The digested wood is also used by the termite.

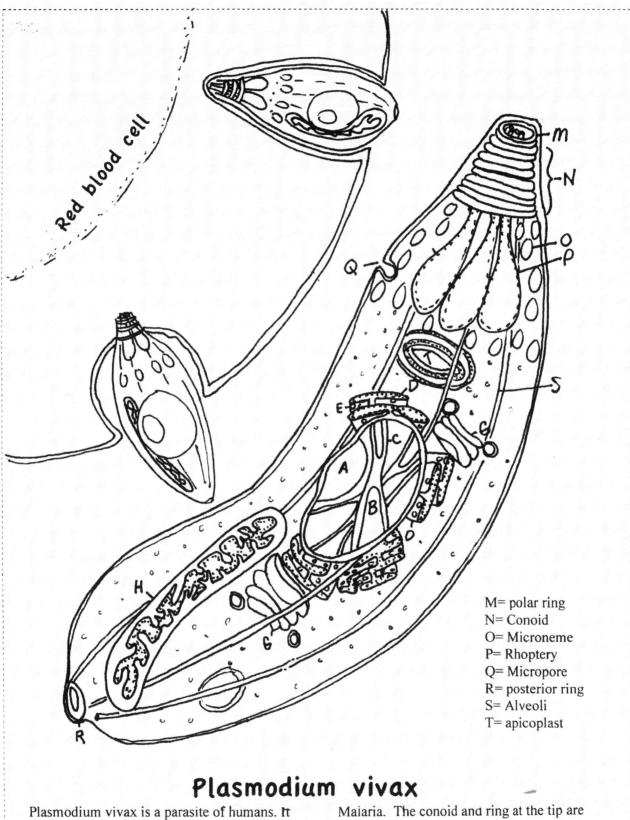

M= polar ring
N= Conoid
O= Microneme
P= Rhoptery
Q= Micropore
R= posterior ring
S= Alveoli
T= apicoplast

Plasmodium vivax

Plasmodium vivax is a parasite of humans. It uses structures in it's "head" to penetrate red blood cells. Rhopteries contain enzymes that help it pentrate the cell. It causes Maiaria. The conoid and ring at the tip are made of tubulin. The alveoli tubules support the membrane so that it is more rigid. The micropore is a mouth-like opening.

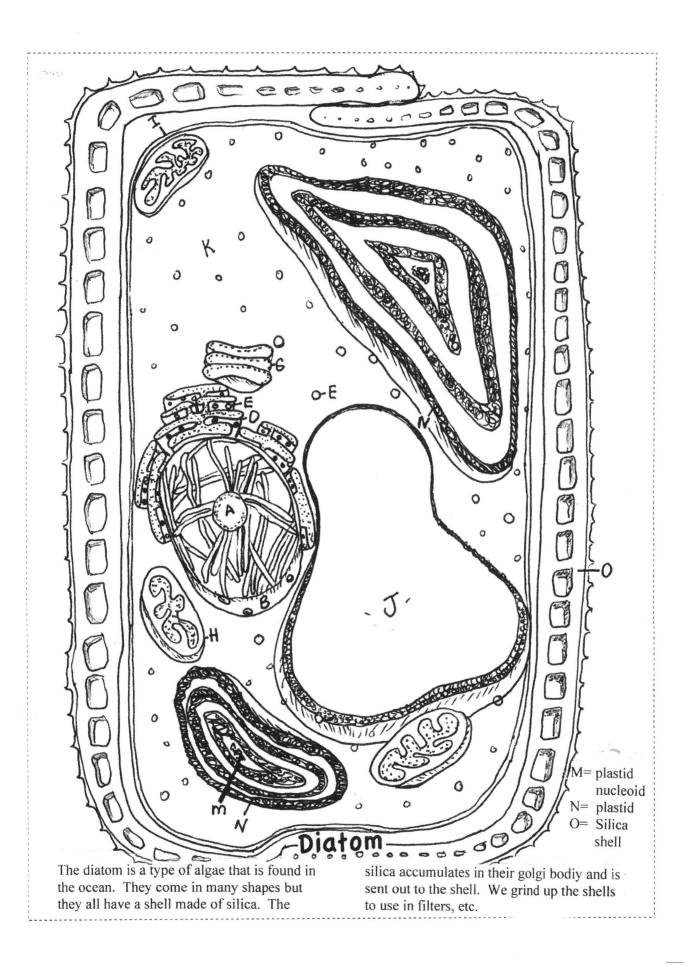

Diatom

M= plastid
    nucleoid
N= plastid
O= Silica
    shell

The diatom is a type of algae that is found in the ocean. They come in many shapes but they all have a shell made of silica. The silica accumulates in their golgi bodiy and is sent out to the shell. We grind up the shells to use in filters, etc.

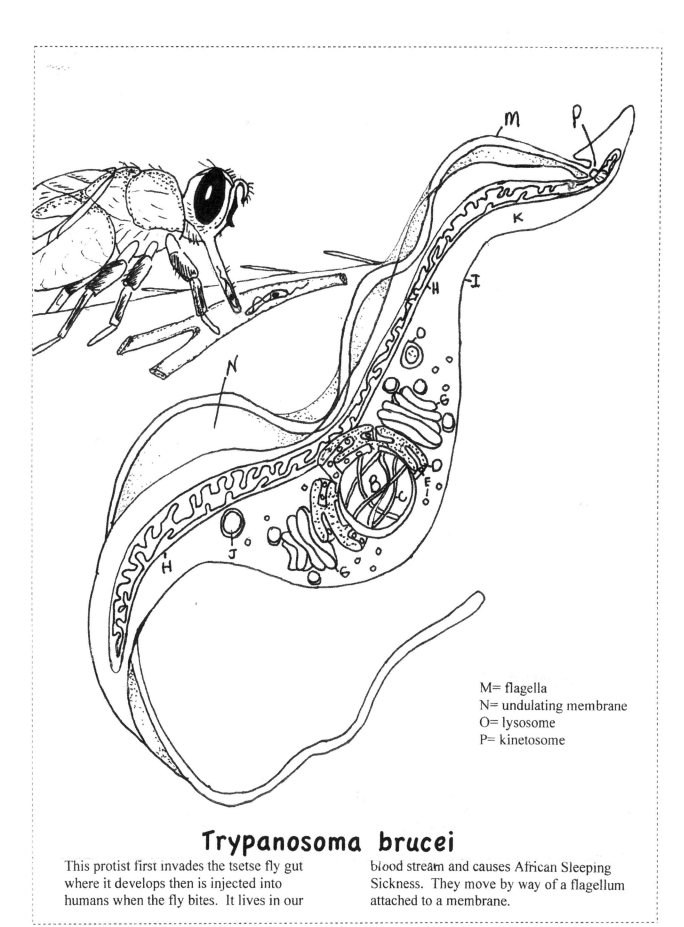

M= flagella
N= undulating membrane
O= lysosome
P= kinetosome

Trypanosoma brucei

This protist first invades the tsetse fly gut where it develops then is injected into humans when the fly bites. It lives in our blood stream and causes African Sleeping Sickness. They move by way of a flagellum attached to a membrane.

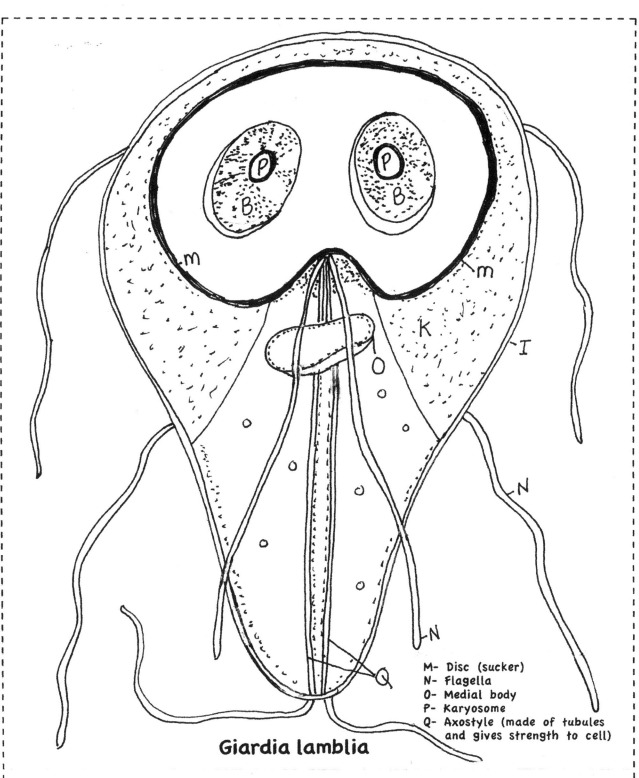

M- Disc (sucker)
N- Flagella
O- Medial body
P- Karyosome
Q- Axostyle (made of tubules and gives strength to cell)

Giardia lamblia

Giardia is a protist that grows in the small intestine causing a disease called Giardiasis. The cell attaches to the lining of the small intestine by a concave disc or "sucker." It gets it's nutrients from food taken in by it's host (human, dog, cow, etc.) There are 4 sets of flagella which help the cell move. There are two nuclei in this cell and each nuclei has a "karyosome" within which contains chromatin. There are no mitochondria, Endoplasmic Reticulum, Golgi bodies or lysosomes in this cell!

Plant Cells

Plant cells all have cell walls made of cellulose. Most plant cells live attached to other plant cells, and thus plants are multicellular like animal cells. Most plant cells have chloroplasts which are green and carry out photosynthesis. This means that in the presence of sunlight, they can take in carbon dioxide and water and produce sugar (glucose) and oxygen. You will notice chloroplasts, mitochondria, cells walls and almost all of the other organelles you have seen in animal cells.

Plants always have to live near the sunlight. In really deep water, plants can't survive because the sunlight cannot reach it. In leaves, plant cells are organized in such a way to take in sunlight and carbon dioxide easily.

Most plant cells have large water vacuoles so that they can store the much needed water for photosynthesis. All the other organelles are pushed to the sides of the cell to make room for the water vacuoles. Even though the choloroplasts are often oval-shaped in plant cells, they sometimes have strange shapes. You will see that the green algae, Spirogyra, has spiral-shaped chloroplasts while the chloroplast of Chlamydomonas is cup-shaped.

The potato cell doesn't have chloroplasts since it is the leaves of the potato plant that carry on photosynthesis. It does, however, contain many *leucoplasts* which store the starch of the potato. These storage organelles stain a dark blue-purple when died with iodine stain. Chlamydomonas also contains starch which are stored in starch granules.

There are around 250,000 known plant species, but only around 150 of them provide most of the food, fiber oil and wood used by humans. Plant diversity is the greatest in the rain forest!

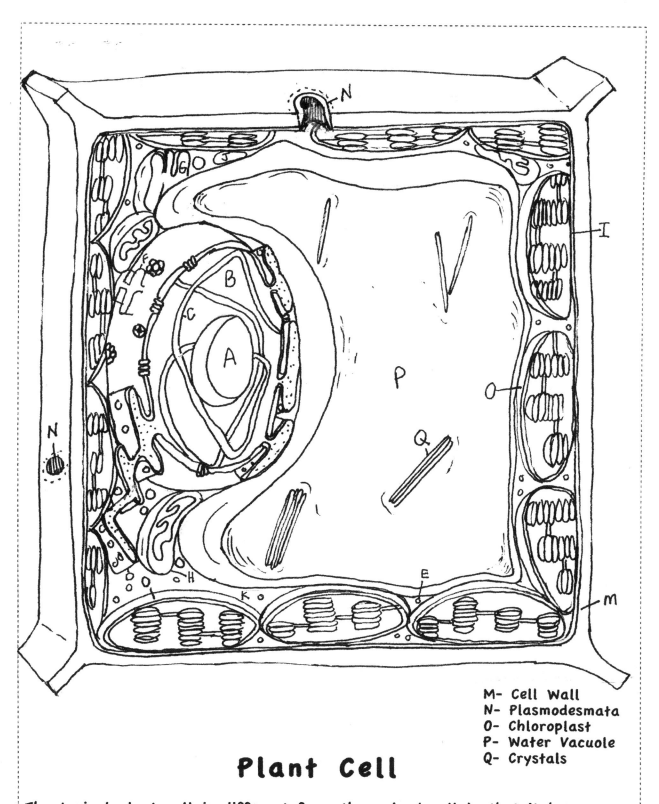

M- Cell Wall
N- Plasmodesmata
O- Chloroplast
P- Water Vacuole
Q- Crystals

Plant Cell

The typical plant cell is different from the animal cell in that it has a thick cell wall to keep the cell stiff and has many chloroplasts which are green and turn sunlight into sugar. It also has a large water vacuole for storage of water for the plant since it needs it to carry out photosynthesis.

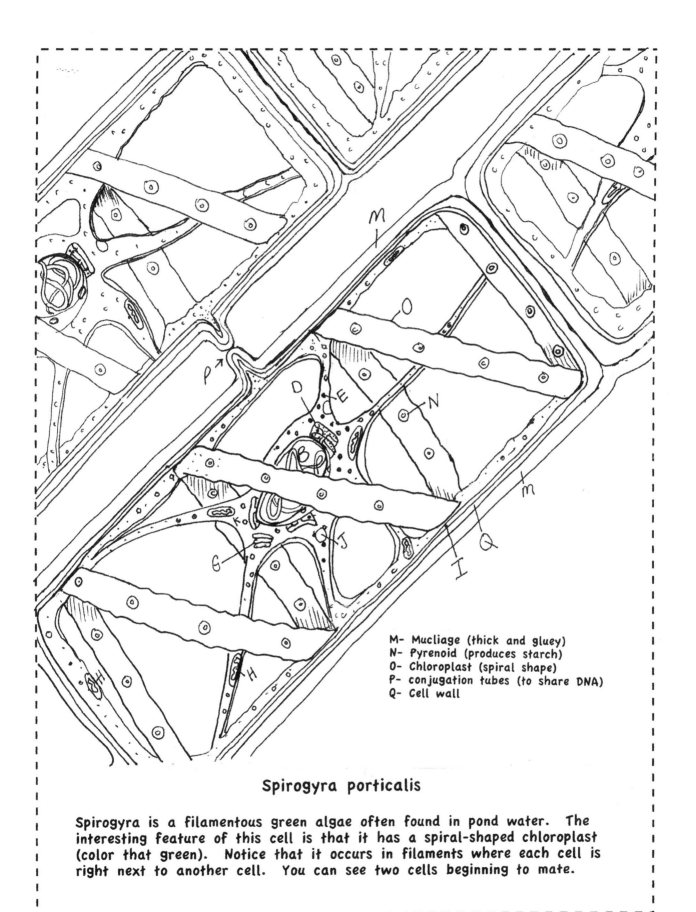

M– Mucliage (thick and gluey)
N– Pyrenoid (produces starch)
O– Chloroplast (spiral shape)
P– conjugation tubes (to share DNA)
Q– Cell wall

Spirogyra porticalis

Spirogyra is a filamentous green algae often found in pond water. The interesting feature of this cell is that it has a spiral-shaped chloroplast (color that green). Notice that it occurs in filaments where each cell is right next to another cell. You can see two cells beginning to mate.

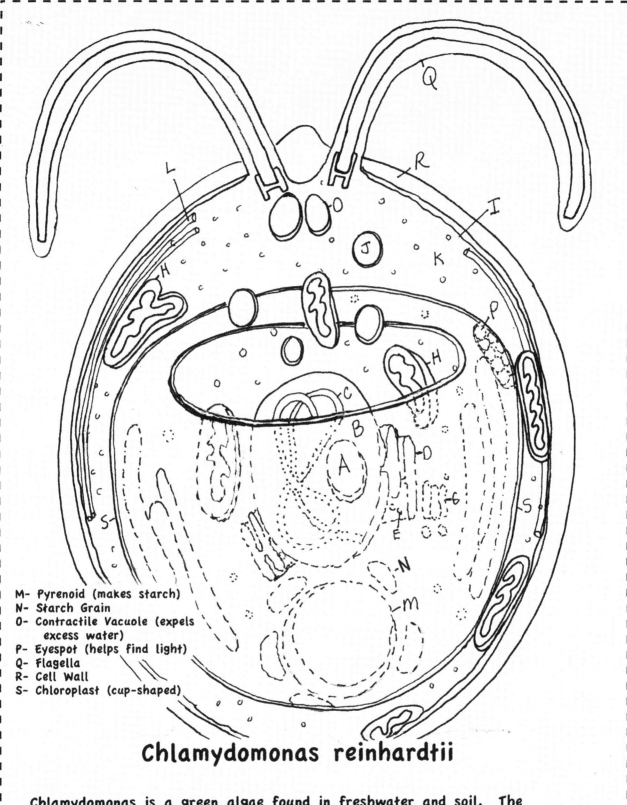

M– Pyrenoid (makes starch)
N– Starch Grain
O– Contractile Vacuole (expels excess water)
P– Eyespot (helps find light)
Q– Flagella
R– Cell Wall
S– Chloroplast (cup-shaped)

Chlamydomonas reinhardtii

Chlamydomonas is a green algae found in freshwater and soil. The mitochondria are often branched and move around the cell. The choloroplast is cup-shaped and surrounds the nucleus and other organelles. It contains a pyrenoid which makes starch and sends it to the starch granules for storage. They have two flagella which helps them move.

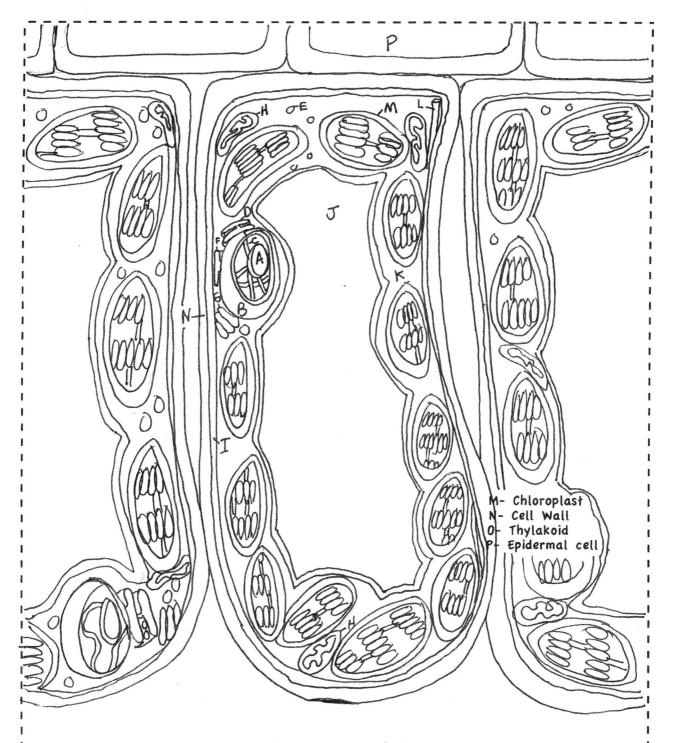

M- Chloroplast
N- Cell Wall
O- Thylakoid
P- Epidermal cell

Leaf (palisade) Cell

The palisade cell is near the top of the leaf closer to the sun. It has many Chloroplasts where photosynthesis takes place. Above these cells are the epidermis of the leaf. Color the chloroplasts green for the "chlorophyll" that is found inside of them. The large vacuole in the center is for water.

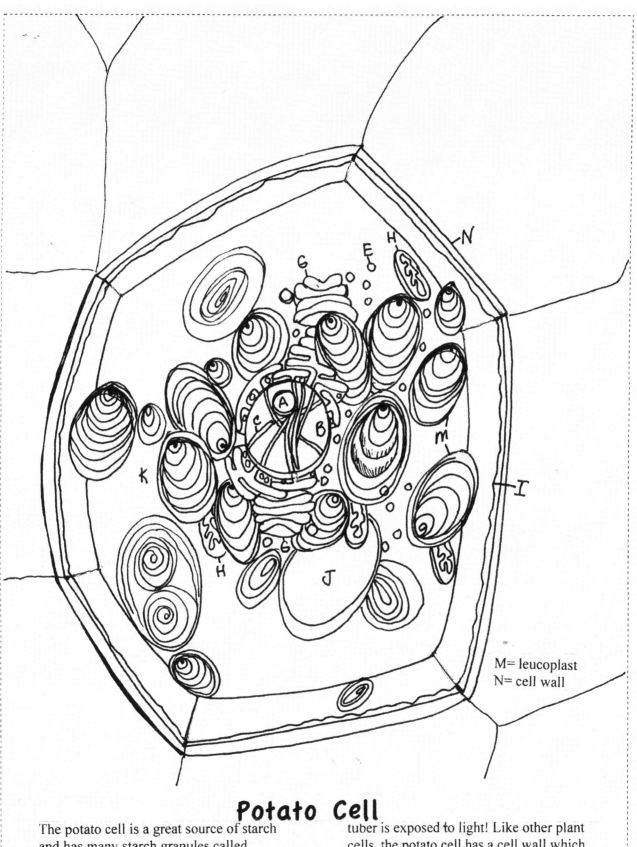

M= leucoplast
N= cell wall

Potato Cell

The potato cell is a great source of starch and has many starch granules called leucoplasts in which to store it. Leukoplasts can turn into chloroplasts when the potato tuber is exposed to light! Like other plant cells, the potato cell has a cell wall which gives it a nice, shape. The potato cell has all the other cell parts, but they are hard to see.

Fungi Cells

Fungi are mostly multicellular, although a really popular fungi in baking, baker's yeast, is unicellular. They have really thick cell walls made of chitin which is the same material that makes up the outer coverings of insects.

Look at the penicillium fungus drawing and notice that it has many filaments made of many cells in a row. Each one of those cells has more than one nucleus. Within each cell, you can see the normal organelles that all Eukaryotes have (animals, plants, protists, fungi) such as the E.R., golgi bodies, etc.

Notice that the yeast cell only has one nucleus and that it reproduces by budding of a new cell. Notice some of the cell organelles going into the new bud to take up residence.

Fungi get their nutrition by absorbing nutrients and frequently secrete digestive enzymes onto their food source to make the food molecules smaller before absorption.

The yeast cell can produce alcohol (ethanol) and carbon dioxide through a process called *fermentation*. It is used in making bread, beer and other foods. Bread dough rises because the carbon dioxide produced as a waste product forms the bubbles that give the dough its structure. Similarly, the carbon dioxide bubbles give beer it's foam.

Fermentation is an ancient process that has been around for a long time. It's not as efficient in making energy as cellular respiration, but has it's advantages too. Fermentation takes place in the cytoplasm while cellular respiration takes place in the mitochondria.

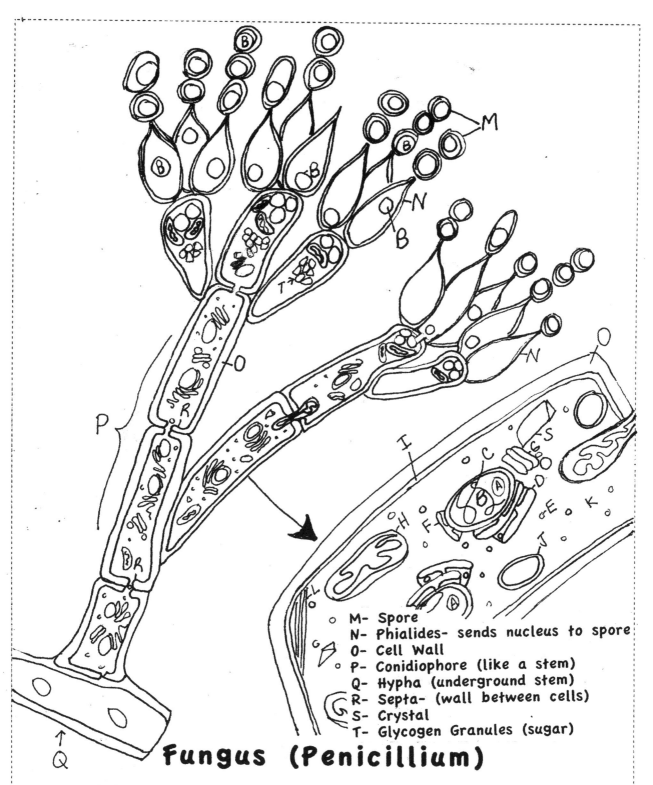

M- Spore
N- Phialides- sends nucleus to spore
O- Cell Wall
P- Conidiophore (like a stem)
Q- Hypha (underground stem)
R- Septa- (wall between cells)
S- Crystal
T- Glycogen Granules (sugar)

Fungus (Penicillium)

The typical fungus cell is similar to a plant cell but lacks chloroplasts.
They have to absorb food from whatever they are growing on! They have a
cell wall that is made of chitin which is like the shells of mollusks. Most
are multicellular except for the yeast. The cells make up a network of
branching tubes called hyphae. Fungi can reproduce by budding or spores.

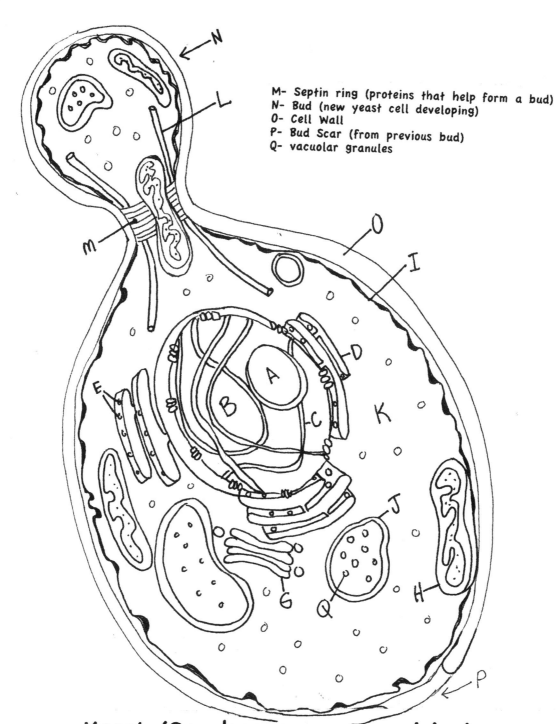

M- Septin ring (proteins that help form a bud)
N- Bud (new yeast cell developing)
O- Cell Wall
P- Bud Scar (from previous bud)
Q- vacuolar granules

Yeast (Saccharomyces cerevisiae)

This one-celled fungus reproduces by budding off a new piece which will grow into a new organism. During a process called fermentation, the yeast produces carbon dioxide and alcohol. It is used in producing beer and bread.

Bacterial Cells

Bacteria are now put into two kingdoms because the two groups are so different. However, all bacteria cells have some things in common. They are unicellular and *prokaryotic,* meaning they lack an envelope around their nucleus and have no membrane bound organelles that we see in the other cells we have looked at (*eukaryotic*). Their ribosomes are different and a bit smaller than eukaryotic ribosomes and they have strange molecules in their cell walls.

The cell walls of bacteria often contain peptidoglycans instead of the chitin or cellulose we see in the other types of cells. Look at the general bacterial cell in the next plate. Notice the fimbriae in the cell wall that helps it adhere to surfaces and the pilus that helps transfer DNA from one bacteria to another.

Bacteria often have flagella that are made of flagellin while the flagella of the eukaryotes is made of microtubules. The bacteria also often have an outer covering called a glycocalyx. The glycocalyx can be a slime layer or a capsule.   The capsule protects them from being eaten by their prey through phagocytosis because it is hard to engulf them when they are slippery. We usually see this capsule in pathogenic (disease-causing) bacteria. If a bacteria has a slime layer, it is for prevention of the leaking off of water and nutrients. Usually a bacteria will only have one of these coatings.

Notice that all bacteria have a cell membrane which is a phospholipid bi-layer. We saw this membrane in all eukaryotic cells too. When the bacteria reproduces by binary fission, this membrane separates in half.

Bacteria cells are the most ancient. Scientists think they showed up around 3 billion years ago while the first eukaryotic cells perhaps showed up a billion years later!

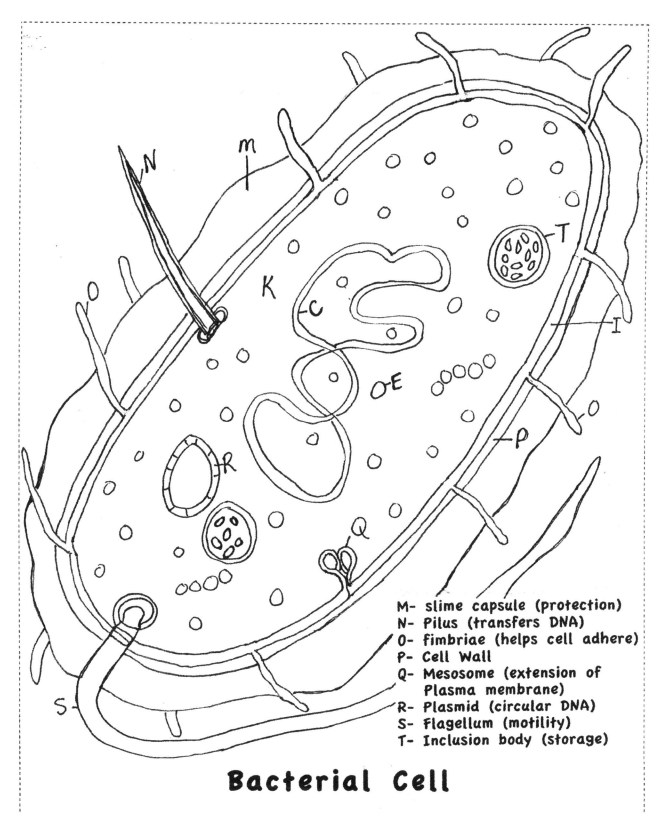

M- slime capsule (protection)
N- Pilus (transfers DNA)
O- fimbriae (helps cell adhere)
P- Cell Wall
Q- Mesosome (extension of
   Plasma membrane)
R- Plasmid (circular DNA)
S- Flagellum (motility)
T- Inclusion body (storage)

Bacterial Cell

The typical bacterial cell is very simple. They have DNA in the middle that is not enclosed in a membrane. They do not have organelles except for ribosomes. They have a cell membrane, wall, and most have a slime capsule for protection. Bacteria usually have one of three shapes: circular, spiral or oblong. Along the surface of some bacteria are pili (pilus-singular) that can help bacteria stick to surfaces and can be a bridge for DNA transfer.

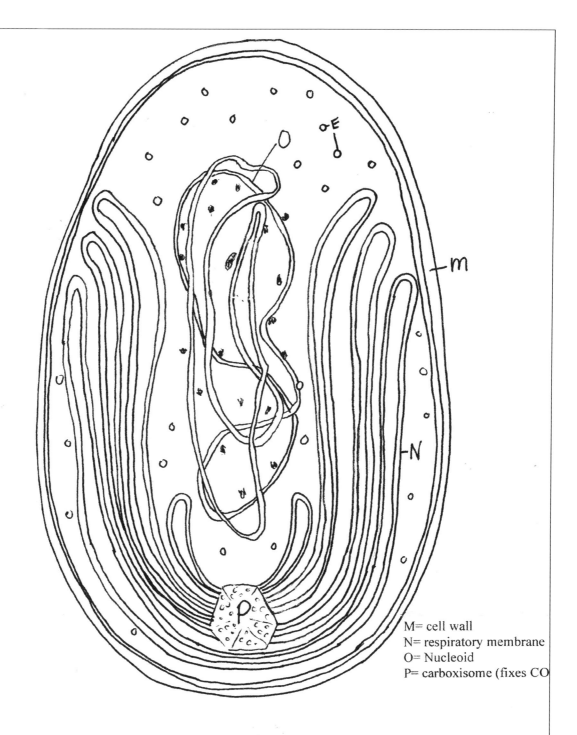

M= cell wall
N= respiratory membrane
O= Nucleoid
P= carboxisome (fixes CO

Nitrobacter winogradski

Nitrobacter is a bacteria that lives in soil or sand. It changes nitrite to nitrate to get energy. It has no cell organelles and it's DNA is contained in the nucleoid. There are ribosomes and cytoplasm, a cell membrane and cell wall.

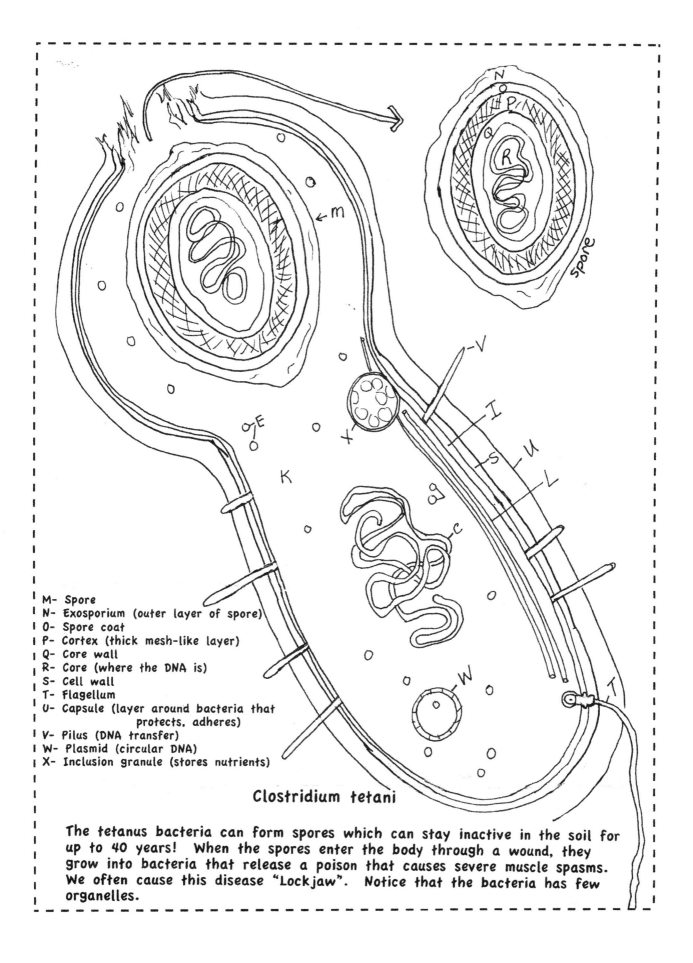

M- Spore
N- Exosporium (outer layer of spore)
O- Spore coat
P- Cortex (thick mesh-like layer)
Q- Core wall
R- Core (where the DNA is)
S- Cell wall
T- Flagellum
U- Capsule (layer around bacteria that
         protects, adheres)
V- Pilus (DNA transfer)
W- Plasmid (circular DNA)
X- Inclusion granule (stores nutrients)

Clostridium tetani

The tetanus bacteria can form spores which can stay inactive in the soil for up to 40 years! When the spores enter the body through a wound, they grow into bacteria that release a poison that causes severe muscle spasms. We often cause this disease "Lockjaw". Notice that the bacteria has few organelles.

Made in the USA
San Bernardino, CA
02 November 2018